YOUR KNOWLEDGE HAS VALUE

- We will publish your bachelor's and
  master's thesis, essays and papers

- Your own eBook and book -
  sold worldwide in all relevant shops

- Earn money with each sale

Upload your text at www.GRIN.com
and publish for free

Bibliographic information published by the German National Library:

The German National Library lists this publication in the National Bibliography;
detailed bibliographic data are available on the Internet at http://dnb.dnb.de .

Imprint:

Copyright © 2015 GRIN Verlag, Open Publishing GmbH
Print and binding: Books on Demand GmbH, Norderstedt Germany
ISBN: 9783656984573

Dennis O'Keeffe

Implementation of Body Worn Cameras for Police Officers

GRIN Publishing

How do you implement an effective body worn camera program for a law enforcement agency?

December 1, 2015

University of Louisville

<u>**Topical Outline**</u>

I. Introduction

A. What is the Need for Body Worn Cameras?

The topic of body worn cameras (BWCs) has gained tremendous attention over the past few years. There have recently been several controversial police shootings around the United States. Those shootings have created an outcry among some communities for more accountability within police work. One effective method of ensuring accountability in police operations is to equip police officers with body cameras. The research in this paper suggests the use of body cameras by the police will make them accountable to the community and make their actions transparent. Transparency and accountability are two words that are often used when there is discussion about body cameras. Law enforcement agencies should be cautious about using the word transparency. Accountability is the more appropriate term when discussing BWCs. Using the word transparent means a law enforcement agency will freely release any video evidence. Very often, police agencies are not in a position to release that video evidence due to the ongoing investigation. Using the word transparency could be the source of contradiction when the public wants to see body camera video. This paper will be a guideline for police agencies to use when starting a BWC program.

I am a current police lieutenant, with 24 years of law enforcement experience, in a large police department (Henrico Police, Virginia) with over 600 sworn officers. In addition to this paper being written to help other law enforcement agencies, it is also being used to complete a Master of Science degree in Criminal Justice from the University of Louisville. I am currently serving as the BWC program manager for my police agency and have successfully deployed over 400 body worn cameras. The BWC research began in 2012 and continues to be an ongoing process. The Henrico Police Division started deploying BWCs in March, 2015 and completed full deployment in December, 2015.

BWCs are a relatively new technology, and limited research is available on this topic. BWCs, by definition, are cameras that can be worn on the body that capture video from the person's point of view (POV) wearing the camera. This paper draws from available research on BWCs and aims to provide law enforcement agencies information, training, and best practices from around the country. This literature review will reveal how to effectively implement a BWC program by focusing on the following key topics: The reason for implementing a BWC program, how to test and evaluate the current camera technology, policy implications, the cost associated with such a program, citizens privacy concerns, retention and storage of data, internal and external stakeholders, and how to continuously evaluate your program to ensure it remains effective and efficient.

B. Definitions

To define the term effectiveness for this paper, we will use the basic definition from the Business Dictionary, The most basic definition of effectiveness is, "The degree to which objectives are achieved and the extent to which targeted problems are solved. In contrast to efficiency, effectiveness is determined without reference to costs and, whereas efficiency means doing the thing right, effectiveness means doing the right thing" (Business Dictionary). It is appropriate that we use this definition of "doing the right thing" because that is the essence of what police agencies are trying to accomplish with the use of BWCs.

Law enforcement is working in a world that is seeing an exponential growth in technology. The citizens in our communities often have better technology than the police. There are an estimated 30 million surveillance cameras in the United States that were used to record 4 billion hours of video per week (Vlahos, 2009). In 2013 alone, there were 120 million smart phones sold in the United States (Hughes, 2014). Each one of these smart phones sold is another video camera in our society. The use of cell phone video, by the citizens recording police actions, is becoming a daily event. The issue with others video recording police action is the fact that it is impossible for them to capture a true representation of what the police officers see and hear, in other words from the officer's point of view. The use of BWCs by the police is an attempt to level the playing field by capturing video from the police officer's point of view. However, even the best quality video has its limitations. This paper will discuss some of the problems and limitations with video evidence.

Law enforcement agencies that wish to implement a BWC program must first identify the internal and external stakeholders. Capturing video of the daily operations of police work will have a major impact on police agencies and those who support police agencies. Consideration must be given to the increased workload for police officers, police supervisors, prosecutors, defense attorneys, IT departments, and administrative personnel that handle freedom of information (FOIA) requests.

It is my goal for this professional paper to help law enforcement agencies implement an effective BWC program. Using the available research and best practices currently being used throughout the country, I plan to explain a thorough process by which agencies can easily adopt a BWC program. This paper will also provide suggestions on how to maintain that BWC program once adopted. There are many facets of BWC technology that are not realized until you are well into the implementation phase. Hopefully, this paper will help agencies avoid those implementation stumbling blocks so they can have an effective and efficient BWC implementation process.

II. Research
A. Literature Review on Body Worn Cameras

A good starting point for any agency wishing to use BWCs is to read *Implementing a Body Worn Camera Program, Recommendations and Lessons Learned,* published by the Police Executive Research Forum (PERF) (Miller and Toliver, 2014). This resource is excellent for police agencies to use because it discusses the benefits of BWCs, privacy considerations, the impact on community relationships, officer concerns, and financial considerations. Agencies will gain a tremendous insight into the various aspects of using BWC after reviewing PERF's recommendations.

One of the major reasons BWCs have gained popularity in American law enforcement is the tremendous attention given to police use of force by the media. In some communities, there is a lack of trust between the citizens and the police. Due to recent controversial police use of force incidents throughout the country, President Barack Obama, on December 18, 2014, signed an Executive Order establishing the Task Force on 21st Century Policing. The mission of the Task Force was to make recommendations to the President about policing practices that can both reduce crime while building community trust between law enforcement and the communities they serve. In March 2015, the Task Force released their Interim Report which outlined six pillars of overarching ideas that can help police build trust in their communities. The first pillar is building trust & legitimacy and that should be the goal for all law enforcement agencies, whether or not they intend on deploying BWCs.

Building trust in our communities and the case for procedural justice is synonymous with effective policing. Procedural justice is the idea that disputes can be resolved in a way that is fair and just for all parties. Procedural justice is a means for attaining legitimacy and is based on four central issues; people want a voice during encounters with the police, everyone wants to be treated with dignity and respect, people want the police to be neutral and transparent in their decision making, and citizens want to know the police have trustworthy motives in their interactions with the community (President's Task Force on 21st Century Policing, 2015).

The American Civil Liberties Union (ACLU) has been very vocal about the use of BWCs and they are calling for consistency among police agencies using BWCs. "For the ACLU, the challenge of on-officer cameras is the tension between their potential to invade privacy and their strong benefit in promoting police accountability" (Stanley, 2013, p. 1). The ACLU is very concerned about the privacy rights of citizens. Law enforcement agencies should also recognize these privacy concerns and mention it in their policy. It is a delicate balance for police to bridge the gap between enforcement action and their citizen's privacy concerns. In October 2015, the ACLU Foundation of Virginia released their model policy along with 15 recommendations that law enforcement agencies should follow when using BWCs. The ACLU of Virginia reviewed 59 local BWC policies in that state and they determined that, as a whole, the policies failed to ensure that BWCs were deployed in a way that promotes transparency, accountability, and trust. They are recommending the state law makers enact legislation to ensure uniformity in police BWC policies among law enforcement agencies.

Chief Doug Middleton (Henrico Police Division, Virginia) brought a unique and innovative training concept to his police department. Chief Middleton partnered the BWC training with Fair and Impartial Policing training. Dr. Lorie Fridell (Professor, University of South Florida) brought her team of trainers to Henrico County and provided her training program of Fair and Impartial Policing. Dr. Fridell teaches police officers how to recognize and overcome implicit biases. She helps police understand that even well intentioned people have biases but to make sure you do not allow those biases to interfere with your decision making with dealing with citizens. As a police officer, when making an arrest the officer should only consider the facts of the situation. All Henrico Police officers receive six hours of Fair and Impartial Policing training and two hours of BWC training. Chief Middleton says that if he had to choose between the BWC and the fair and impartial training he would choose the Fair and Impartial Policing training over the BWC because of the importance of that training in recognizing implicit biases.

There has been a tremendous increase in attention, by the media and public, about the use of body worn cameras by police. Despite this increase in attention, there has been very little research on body worn cameras and little to no research on the officer perceptions of those cameras. In 2014, Jennings, Fridell, and Lynch conducted a randomized experiment, evaluating the impact of body worn cameras, by studying baseline data collected from surveys given to Orlando Police officers. The survey results show that most officers are supportive of the idea and use of body worn cameras. They acknowledge that body worn cameras will improve behavior both on the part of the citizen and officer alike (Jennings, Fridell, & Lynch, 2014).

It is important for law enforcement to get buy-in from their officers before developing a BWC program. Earlier this year, Drover & Ariel explained the process of implementing a randomized controlled trial for using body worn video in a large British police agency. Even though the purpose of their trial was to measure the effects of video on use of force and complaints, they also acknowledged the fact that they first must ensure they gained the support of the test group. Their paper explains the procedures used to implement this trial which gained the support of the users of body worn cameras. The content of this paper can help police leaders with understanding evidence-based testing, help them implement change in their agency, and help manage their police operations. The techniques used in this trial can be used by other police agencies to effectively implement their body worn camera program (Drover, & Ariel, 2015).

B. Use of Force – Police vs. Citizen Perceptions

The use of force by police against citizens is a global concern of both academics and practitioners. Police use of force against citizens is one of the most controversial aspects of police work. Do BWCs reduce the number of use of force incidents and/or citizens' complaints against the police? Ariel, Farrar, and Sutherland used a randomized controlled trial to empirically test the use of BWCs by measuring the effect videoing has on police-citizen encounters regarding use of force and complaints. This randomized controlled trial covered a 12 month period whereby they assigned officers to either a "controlled-shift" who did not receive a camera, or an "experimental-shift" whose officers received a camera. The results from this experiment showed that both use of force and complaints were significantly reduced in the group with cameras (Ariel, Farrar, & Sutherland, 2015).

Police agencies should be very cautious about implementing a body worn camera program without first doing thorough research and a trial run with a few cameras. We recognize that BWC technology is new technology, there is very little research, and some of the research done so far has significant limitations. There are three things that should be accomplished during the research and planning phase of implementing a body worn camera program. First, agencies should define their purpose of implementing their camera program and clearly explain and define that purpose to all the stakeholders. Second, agencies should choose the camera hardware specifications that are needed by their agency. Once they know what specifications the camera must have, they can then test and evaluate camera vendors who meet those specifications. Third, does the camera company they choose have the support services needed to make your program successful? The ideas outlined will give valuable support to law enforcement agencies when explaining how to implement an effective body worn camera program (Dietzen, 2015).

There are specific strategies involved with implementing BWCs and they are the future for law enforcement. BWCs provide many benefits, but there are also several myths about the use of BWCs. Some of the myths are objectivity and perception about how the video is interpreted when viewed by citizens. Another myth is that BWCs will not be the "silver bullet" for improving police-citizen relations. There are several Department of Justice (DOJ) and Police Executive Research Forum (PERF) reports that outline the police use of body worn cameras. Geoghegan discussed the increased use of body worn cameras, and explains the Los Angeles

County Sheriff's Department's pilot program for implementing body cameras. That article provides some insight about looking to the future with this body worn camera technology by thinking critically about the issues surrounding BWCs and to continue improving upon your policy and best practices (Geoghegan, 2015).

Agencies that choose to use BWCs must understand that cameras will not be the panacea that will automatically build trust between their communities and their police force. An officer's BWC could malfunction or they could forget to start the recording, which would be looked upon as deceitful by the community. There are several limitations to BWC video and police agencies must have personnel who are knowledgeable about these limitations so they can explain these limitations to citizens, lawyers, judges, and juries. Dr. Bill Lewinski, Executive Director of the *Force Science Institute,* lists 10 limitations of BWC that agencies should be aware of for their protection:

> **1. A camera doesn't follow your eyes or see as they see** – an officer could glance away from the BWC field of view or miss things the BWC may view. The BWC also cannot capture the human aspect of real life. Officers experience physiological and psychological effects while under stress that cannot be captured on video.
>
> **2. Some important danger cues can't be recorded** – if an officer has their hands on someone, such as during an arrest, they can feel "tactile cues such as resistive tension" according to Dr. Lewinski. This is the very minute muscle movement by someone beginning to resist or assault. When the officer reacts to this "resistive tension" by using force, it often does not match what is portrayed on the video.
>
> **3. Camera speed differs from the speed of life** – reactions times must be understood during real life situations. If it takes the human body a half a second or more to respond to a perceived threat, then the same is true on the opposite end. Once there is no longer a threat, it takes the human body a half a second or more to stop responding to that threat.
>
> **4. A camera may see better than you do in low light** – it must be understood that the BWC can never duplicate the human eye, when reviewing BWC video some objects may have better clarity than what the officer saw. Transitioning from a dark-light environment or vise-versa, could cause distortions or black out images in the video.

5. Your body may block the view – very often a part of the officer's body may block the BWC view. The various mounting options used by officers make this an issue that may arise. An officer drawing and pointing a firearm or TASER will often block the view in a chest mounted BWC.

6. A camera only records in 2-D – when reviewing BWC video it is often difficult to judge distances and depth that can be judged by the third dimension of the human eye. Reviewers of BWC video must understand this and be able to explain this to citizens.

7. The absence of sophisticated time-stamping may prove critical – BWC video that is captured of a controversial shooting or use of force incident must be analyzed by slowing down the video to one-hundredths of a second or less. There are computer programs that can add the necessary time stamps to more thoroughly analyze the video and Dr. Lewinski strongly recommends this when investigating the incident.

8. One camera may not be enough – the angle, lighting conditions, and several other factors can affect the perception of the video. One officer's BWC video may show a perceived improper use of force, when another BWC video of the same incident will show a perceived justified use of force. One great example is the use of instant replay in professional sports, they will often need several video angles to get the call right. Very often, none of the many cameras will show what truly happened.

9. A camera encourages second-guessing – there are many people who want to second-guess an officer's action after a controversial shooting or use of force incident and BWC gives them their platform for second-guessing. These would-be second-guessers should be reminded of the Supreme Court's 1989 decision in *Graham v. Connor,* "An officer's decisions in tense, uncertain, and rapidly evolving situations are not to be judged with the 20/20 vision of hindsight."

10. A camera can never replace a thorough investigation – video evidence is just another small piece of evidence of the many that must be collected. Officer, victim, suspect, and witness statements must still be recorded and vetted. Physical evidence must be collected and analyzed and by using all available evidence can we more accurately re-create what happened. The human factors of reaction times must also be taken into consideration during the investigation process.

Your police agency must have personnel in the BWC program that are knowledgeable about the limitations of BWC video and the human physiology of reaction times. Your police agency must also be able to explain these things to any audience.

Use of force by the police against citizens and suspects will always be a topic of controversy and debate. The reason for this controversy is that everyone has varying perspectives about how, when, and the amount of force that should be used by the police. BWCs will now give yet another perspective whenever force is used by the police. The landmark case governing police use of force was *Graham v. Connor* 490 U.S. 386 (1989). In this case, the Supreme Court ruled that any force used by the police must be "objectively reasonable" and that before using any force, the police must consider three things. The police must consider the severity of the crime the suspect allegedly committed, whether the suspect is trying to cause injury or harm to the officer or another person, and whether the suspect is trying to resist arrest or flee, before using force by the police. With regard to what is considered reasonable, the court ruled that the reasonableness of the amount and type of force used must be judged from the perspective of a reasonable officer on the scene, rather than the 20/20 vision of hindsight. How we as individual citizens define and determine what is reasonable will be a debate that continues for a long time. It is very difficult to determine and define reasonableness (Alpert & Smith, 1994). Although the Supreme Court case of *Graham v. Connor* gave clearer guidance to police, there is still much debate over what is reasonable use of force by police. Police agencies deploying BWCs must be prepared to educate their citizens about why each use of force was reasonable in their eyes.

There are varying perspectives about police use of force among citizens, the media, special interest groups, religious leaders, civic groups, and law enforcement. The use of BWCs will give us another perspective. Trying to close this gap of varying perspectives is a daunting task, but not impossible. We all must make an effort to see through each other's eyes. "Social theory can serve many functions in the public policy arena. Two of the most important in the realm of crime and justice are: (1) guiding the actions of criminal justice agencies and personnel; and (2) explaining to members of the public how and why agencies and personnel act the way they do" (Klinger, 2005, p. 1). All law enforcement officers must get out into the community and interact with their citizens so they can explain why they do the things they do. The community must also interact with officers and explain their views and needs. This is certainly true when choosing to deploy BWCs. Police agencies must educate their citizens on the strengths, weaknesses, and limitations of body cameras.

There are several model policies and training recommendations available to police about how to train officers in using force. The Police Executive Research Forum (PERF), International Association of Chiefs of Police (IACP), and the Department of Justice (DOJ) have all written model policies and made recommendations for police agencies about the use of force. The consensus among these organizations is that police agencies should have sound policies governing the use of force, and those policies should not contain ambiguous language. Lee, et al. (2010) conducted an examination of police use of force utilizing police training and neighborhood contextual factors. "This study showed that violent crime rate and unemployment rate were significant factors as the neighborhood contextual variables" (Lee, Jang, Yun, Lim, & Tushaus, 2010, p. 3). His study also showed that in-service training for police was a significant "organizational-level" factor that explained the levels of force used by police. There is no question that police training on use of force varies among police agencies. It is a good idea to incorporate use of force training with your BWC training so officers have an understanding about the issues we face with both topics. Those agencies that have sound training practices and policies are less likely to use improper or excessive force.

Police use of force against citizens often is the result of a lack of understanding between the police and the citizen. Communication and verbal diffusing techniques can prevent many uses of force encounters. Often, citizens are unaware that the officers are required by state law or department policy to take enforcement actions, and this often leads to conflict between the two. Rojek, Alpert, & Smith conducted a study that examined officer and citizen accounts of police use of force incidents. Their goal was to understand the dynamics of why police-citizen encounters result in force being used. "Our research uncovered situations where officers justified their actions to maintain their authority. Officer respondents consistently sought to explain escalating combative behavior as based on a threat to themselves and/or the safety of the public and simultaneously transformed their perception of citizens into suspects" (Rojek, Alpert, & Smith 2012, p. 322). The data from their study show that the citizen and the officer often focus on different issues during their interaction and, "justify their behavior by the identification and maintenance of their self-prescribed roles" (Rojek, Alpert, & Smith 2012, p. 301). This research showed that the officers would justify their actions to maintain their authority. The officers would explain that the escalating combative behavior was based on threats they perceived against them or others within the public. The citizens upon whom the force was used reported feeling victimized by the officer and they felt there was no way to negotiate their way out of the encounter with the police. Such "accounts" by social actors are similar to Sykes and Matza's (1957) Techniques of Neutralization, specifically the "denial of responsibility" and "condemnation of the condemners" on the part of citizens (Rojek, Alpert, & Smith 2012).

The media can be both a tremendous asset, and a detriment to police. During any highly publicized police-citizen encounter, especially when force is used, the media will be there trying to be the first to get the story out. The media is quick to interview "eye-witnesses" and will broadcast their accounts of the incident without taking the time to substantiate the information. The public will often view these stories and believe what they are hearing from the media is valid information, when often it is inaccurate information. The police are often restricted in the information they release because they do not want to jeopardize the criminal investigation. The police are in a position where they have to balance the public's right to know vs. the privacy rights of those involved. The impact the media has on public perceptions about police-citizen encounters has been well documented (Beckett & Sasson, 2000; Flanagan & Vaughn, 1995;

Ford, 1997; Sigler & Curry, 1992; Tuch & Weitzer, 1997). The public knows very little about the criminal justice system, yet criminal justice policies are often shaped by public opinion (Roberts, 1992). It is important for law enforcement agencies to reach out to the media outlets in their areas and try to build relationships in an effort to bridge the gap between the lack of understanding between citizens and police about police work. "Studies examining the media and its effects on public attitudes towards crime and the criminal justice system have found that public opinion is greatly influenced by the media" (Jefferis, Butcher, & Hanley 2011, p. 83). Maxson et al. (2003) found that 35% of the subjects in their sample had based their knowledge of the police solely on media accounts. Well-publicized incidents of police violence had a negative and prolonged effect on public perception of the police (Weitzer, 2002).

One of the most often highly publicized use of force tool used by law enforcement is the TASER. Video cameras can also be attached to the TASER. The TASER CAM and the BWC will both give you varying perspectives of the incident. The TASER is an electronic control device that is designed to be used as an alternative subduing method for law enforcement when traditional control tactics have proven, or are likely to be, ineffective. The device emits electrical energy that temporarily overcomes the human neuromuscular system. Over the years, there have been many descriptive names for the TASER; Electronic Control Weapon (ECW), Conducted Energy Device (CED), and the currently used name of Conducted Electrical Weapon (CEW).

In March of 2011, the Police Executive Research Forum (PERF) released their guidelines that law enforcement should follow when using CEWs. The guidelines were organized into six categories: agency policy, training, using the CEW, medical considerations, reporting and accountability, and public information & community relations. In an effort to help educate those outside of law enforcement, the following recommendations were made, "Law enforcement agencies should conduct neighborhood programs that focus on CEW awareness training, which should be part of any citizen's training academy program. Agency's public information officers should receive extensive training on CEWs so they can better inform the media and the public about the weapon. Members of the media should be briefed on agency's policies and use of CEWs. CEW awareness should extend to law enforcement partners such as local medical personnel, citizen review boards, medical examiners, mental health professionals, judges, and local prosecutors" (PERF, 2011, p. 24).

Research shows that CEWs can reduce injuries to officers and suspects. Anytime law enforcement has to use force, there can be problems, but the use of CEWs is a reasonable use of force by police when used in the right circumstances. "When left uncontrolled, any use of force can be dangerous, but when managed in a system that includes strong policy, training, supervision and accountability, the use of a CEW can be a great asset to the police and the community by more effectively reaching the goal of using the least amount of force necessary to control unruly suspects, while minimizing injuries to officers and suspects" (Alpert & Dunham 2010, p. 255).

One of the most often misunderstood citizen perceptions about the use of force by police is that of reaction times. Dr. Bill Lewinski has done a tremendous amount of research about the science behind the reaction times of human beings. He teaches an intense and in-depth 40-hour course (Force Science Certification Course) that teaches the attendees the science behind the reactions times, and how we all react to threats or perceived threats. Citizens will often view video of a police shooting and they do not understand why the officers continue to shoot when the threat is no longer present, why police sometimes shoot a suspect in the back, and why the officer felt threatened when the video does not appear threatening. With regard to reactions times, Dr. Lewinski explains that it takes the human body anywhere from .5 to 3.0 seconds to respond to a perceived threat. Conversely, the same is true when you realize the threat is gone. It takes the human body .5 to 3.0 seconds to stop reacting to a perceived threat. If an officer is firing a weapon at a suspect, the officer cannot immediately stop firing the instant the threat is gone. The same is true when a suspect gets shot in the back. If an officer has an immediate threat and makes the decision to use deadly force and the suspect turns after the officer has made the decision to fire then the suspect will often get hit in the back. The BWCs will inevitably capture video of these use of force incidents by officers against citizens. It is important for your police agency to have staff in place that can explain these reactions times to those viewing the video (Lewinski, Hudson, & Dysterheft, 2014).

Dr. Lewinski also explains the "looming" effect during his force science training. When a police officer fires their weapon at an automobile coming towards them, they could suffer from the "looming" effect. This occurs when the officer finds themselves directly in the path of the oncoming automobile. According to Dr. Lewinski, the "looming" effect is when the perceived speed of the vehicle is much greater when you are directly in the vehicles path. When reviewing the video of this event, or if you were present at the incident but at an offset angle from the vehicles path, you would not perceive a threat. It is important for police agencies to educate the public, and the media, about the science behind human behavior and reaction times.

C. Test and Evaluate (T&E) Available Body Worn Cameras

Once your agency has conducted a thorough literature review and research on BWCs it is now time to test and evaluate (T&E) the available BWCs on the market. Most BWC vendors will send you one or more BWCs to T&E upon request. If the vendor also provides storage options, they will also provide access to that storage solution for you to test. When testing BWCs, it is a good idea to choose officers who are hard workers, conscientious, and make a lot of arrests. The hardest working officers will provide the best real test for the BWCs you are evaluating. You should create a checklist of items that you want to know about regarding the function of the

BWC. Creating a checklist will allow all officers to consistently evaluate each BWC being

tested. Some of the most popular BWC companies are:

> TASER International, Inc.
> Digital Ally, Inc.
> VIEVU
> WatchGuard Video
> Wolfcom
> Panasonic
> Reveal

In addition to the vendors listed here, there are dozens more out there for you to test. The key is

to find a BWC vendor that suits the needs of your agency. T&E as many BWCs as you possibly

can to get a true idea of the quality of the products on the market. You may find that some

vendors will require you to purchase their BWC if you wish to test it. Your agency will have to

decide if that is a viable option for you, but most of the industry leaders will give you a BWC to

test at no charge. Thoroughly testing the BWC you wish to purchase will save you a great deal of

time, energy, and money in the long-run because you will know the limitations and what to

expect from that BWC.

D. Data Storage Options

There are basically two types of storage options for your BWC video, cloud storage or in-

house or local storage options. If you have a small police agency (fewer than 25 officers) you can

probably use your current local storage options. Very often, local IT workers will argue to keep

your data on your servers, or your local network. This is usually because it is job security for

those IT workers. Cloud technology makes it easier to manage data and you will always have a

need for those IT employees. Larger agencies should consider cloud storage options. Any

agency, large or small, will probably find cloud storage options are far less expensive and easier

to manage for their BWC video. The Federal Cloud Computing Strategy recommends that all government agencies use cloud first technology when possible. This Federal Cloud Computing

Strategy is designed to:

> Articulate the benefits, considerations, and trade-offs of cloud computing
> Provide a decision framework and case examples to support agencies in migrating
  towards cloud computing
> Highlight cloud computing implementation resources
> Identify Federal Government activities and roles and responsibilities for
  catalyzing cloud adoption

The federal strategy will allow us to perform more effective and efficient as an organization

which in turn will allow us to better serve our citizens. We have a responsibility, as a

government agency, to move towards this cloud first technology as quickly as possible to gain

the benefits of cost and innovation the cloud provides (Kundra, 2011).

III. Tasks to Accomplish Before Receiving Your Body Worn Cameras
A. Choosing a Program Manager and Team

Choosing a BWC program manager and team members is vital to the success of your

BWC program. Regardless of the size of your police agency, the people that are chosen to lead

this program are the ones who will make this a great program or a marginal program. Ideally, the

person you put in charge should have several honed skills to include, leadership skills, effective

teaching skills, and a good working knowledge of computers, camera systems, and video

management. There are many attitude characteristics a truly great leader will display including,

loyalty, appreciation of subordinates, authority with responsibility and accountability,

enthusiasm, and integrity (Ponder, 2001). Your organization will also have a much easier time

with implementing a BWC program if the project manager is someone your police officers

already know and trust. Choose your BWC team carefully because your BWC program will be

very effective, or it could struggle, depending upon the people you choose.

B. Consult the Internal and External Stakeholders

Your law enforcement agency must consider all internal and external stakeholders your BWC program will affect, and you must consult with each of these stakeholders. Many of these stakeholders will often be unnoticed until your BWC program increases their workload. The first stakeholder to consider is the officer, the end-user of the BWC. Your agency should include officers in the testing, purchasing process, and policy creation of your BWC program. Patrol supervisors and the command staff should also have input in the decision making stages of the BWC policy creation. The information technology (IT) workers in your police department play a vital role in the development of your BWC program. The IT staff must be consulted to ensure your buildings infrastructure can accommodate the increased bandwidth and storage that comes with the BWC videos. The judges, prosecutors, and defense lawyers must all be consulted before bringing BWCs to your agency. BWC video must be shared with your prosecutors and defense lawyers and this video will increase the workloads of both offices. The judges must be consulted because now that we have this video, we will need to come up with a method to play these videos in court when needed. The last, but certainly not the least, stakeholders are the citizens in your community. This is the one stakeholder that often goes unnoticed when law enforcement agencies are moving towards a BWC program. The decision of whether or not to bring BWCs to any community should be decided in collaboration with the elected officials, law enforcement, and the citizens of that community. If then, the consensus is to bring the BWC to that jurisdiction, the community should have a say in the policy creation. If you reach out to all these stakeholders before purchasing your BWCs, your agency will be off to a great start in implementing a very effective BWC program.

C. Purchasing Process

The purchasing process will have various guidelines you must follow depending on your states purchasing laws. Law enforcement agencies must work closely with their jurisdictions purchasing office to acquire BWCs. In Virginia, government agencies wishing to purchase goods or services over $15,000.00 must send out Requests for Proposals (RFP) or Invitation for Bids (IFB) to comply with the states purchasing laws. The police agency will be responsible for writing the specifications for the BWC you want and the scope of services you are seeking. The following is an example RFP for BWCs and video storage:

<u>SCOPE OF SERVICES</u>: The Successful Offeror shall provide all equipment, labor, materials and supervision for the following:

A. Requirements for body worn camera: The camera must have the following characteristics:

1. Meets MIL-STD 810F Method 506.4 Procedure I – Rain and Blowing Rain.
2. Has configurable bit rate and audible settings.
3. Has low light capability, <=.1 lux.
4. Must be able to record for at least four hours per activation to allow for lengthy interviews and investigations.
5. Video capture must reflect the point-of-view of the officer, both audio and video.
6. Must have an angle of view of at least seventy degrees. The view must not be obstructed when an officer has a firearm out and in the ready position. A head mountable camera is preferred.
7. Multiple mounting options to accommodate varying field situations are preferred.
8. Must have a frame rate of at least 30 frames per second at required resolution.
9. Must have a video resolution of 640 x 480 or better.
10. Must have a minimum of 8GB of internal storage memory.
11. Must contain a buffer that captures a minimum of 30 seconds of video which is saved upon activation.
12. In addition to record and buffer, system must offer a privacy mode where no recording is taking place.
13. Must have a switch or push button that is easily accessible to activate.
14. Shall produce an audible tone periodically to denote recording. The camera must include a volume control for the operator to lower when necessary.
15. Must have a user option which allows illuminated controls or indicators to be extinguished during a tactical/darkness situation.
16. Must have a rechargeable battery that;
    a. Has a minimum of 12 hours of standby time;
    b. May be easily changed by the user without the use of tools;
    c. Has the ability to be charged in the field.

17. Must work with system that prevents user from deleting or editing the original video file.
18. Must be lightweight. Systems with less weight will be preferred.
19. Must have the ability for users to review video in the field.
20. Must be durable. Offeror should describe the durability of the system equipment.

B. Requirements of Storage Solution:

1. The Successful Offeror shall provide a cloud based data storage system with the capability of organizing and managing the stored video. In addition to a user identifier and date/time when the video was recorded, it should permit the storage of a minimum of 5 meta-data fields defined by the County during the initial implementation planning.
2. The Successful Offeror shall provide cloud based data storage and management software for 5 years. The County shall have the right to terminate data storage, training, and hardware and software maintenance services after the first year of the contract by notifying the Successful Offeror 90 days prior to the contract's anniversary.
3. The data storage system must meet all FedRamp requirements for cloud storage and the cryptographic modules must meet the FIP 140-2 standards for data at rest and in transit.
4. From the date of storage, the system must provide a minimum of 90 days storage and retrieval for all video. The system shall allow the storage for at least one year of up to 25% of all video uploads, when identified for longer storage by the County, as well as storage for at least two years of up to 5% of video uploads as identified by the County.
5. The system must allow for automatic video transfer from cameras to cloud-based storage via a multi-charging/transferring docking station with USB cables from individual computer running Windows 7 or 8.
6. The software management tools must at a minimum provide:
    a. Ability to manage the users, roles and permissions to upload, view, download and other applicable operations.
    b. Ability to easily export/download video in a non-proprietary, watermarked, read-only file format suitable for evidentiary/court purposes. File(s) should be viewable using readily available software such as, but not limited to, Microsoft's Media Player and Adobe Flash.
    c. Ability to create video clips extracts from video files.
    d. Availability of an audit trail for every video which tracks all user activity.
    e. Ability to set retention rules and allow administrators to delegate/purge files based upon those rules.
    f. Ability to track and assign all devices.
7. The Successful Offeror must provide customer service support to assist with all software issues within one business day.
8. The Successful Offeror must provide a method using best practices in use when storage services end for transferring video, metadata and data definition to the County or another vendor.
9. The Successful Offeror shall have personnel available to provide expert witness testimony in court regarding the system.

C. Training

The Successful Offeror shall provide training to the County for all features and services of the system. Offerors shall describe in detail their training capability and any costs. Offerors shall provide their method of training, *i.e.*, on-site, web-based.

D. Implementation

1. Offerors must be able to deliver equipment, training and cloud-based storage in accordance with the schedule in paragraph 2 below.
2. The initial order will be for 36 cameras to be delivered before the end of calendar year 2014. The County intends to purchase 364 additional cameras during calendar year 2015 according to the following schedule: 1st quarter - 64 units; 2nd quarter 100 units, 3rd quarter - 100 units and 4th quarter - 100 units.
3. The contract pricing per camera unit shall include the battery and mounting and shall apply to all purchases before December 31, 2015.
4. The County may also wish to purchase additional cameras during the contract period.

E. Warranty

1. The Successful Offeror shall warrant operation of all software and hardware products for a period of 12 months from the date of final acceptance by the County.
2. Successful Offeror shall provide five years of maintenance and support and include its cost as a separate line item in the Pricing Schedule. The first year of maintenance and support will be provided as part of the initial cost of the equipment. All yearly maintenance and support fees for subsequent years shall be provided at a fixed per year price. The County may terminate yearly maintenance and support by giving written notice to the Successful Offeror at least 90 days before the annual contract anniversary.

Henrico County, VA RFP # 14-9671-10YD

This is a good example for any law enforcement agency to use when creating a RFP or IFB. The key is to involve the purchasing department from your jurisdiction to ensure you are complying with your states purchasing laws. When discussing the BWC key technical specifications with your vendors, there are several key items you want in a BWC, such as;

- Video resolution
- File metadata (date/time stamp, officer ID or unit number, etc.)
- Low-light settings
- Safeguarding of the video against editing, tampering, access, etc.
- Pre-event recording (usually 30 seconds)
- Battery life/performance
- Operational cycle (officer picks up camera at start of shift, downloads at end of shift, places in charger, etc.)

> ➢ Video management

(Bolton, 2015)

Your agency should ensure the vendors responding to your RFP or IFB specifically address these key items in their response.

D. Policy Considerations

The last thing you must do before actually putting the BWCs in the field is to draft a policy for their use. It is a big mistake for any law enforcement agency to put BWCs in the field without first having a policy governing their use. There are many model policies available for review and use, including PERF, IACP, ACLU of Virginia, and Virginia Department of Criminal Justice Services (DCJS). Your agency should review all available model policies, and then construct a policy that works for your agency. Your BWC policy should contain, at minimum, sections on purpose, policy, regulation, procedures on when to start and stop recording, restrictions on use, evidence collection, access & retention, and supervisor responsibilities. Once your draft policy is written, you should make an effort to share your policy, and seek input, from all the stakeholders, including your community. There is no better way to show you are being transparent and accountable than to share your policy and ask for feedback from these stakeholders.

IV. Issuing Body Worn Cameras
A. Which Officers Should be Issued Body Worn Cameras

Your police agency will need to determine which officers will be issued BWCs. Most law enforcement agencies issue BWCs to uniformed officers that have the potential to interact, or take enforcement action, with the public. The bulk of your BWCs will be issued to patrol officers because the patrol section is the backbone of any police organization. Your agency will have to decide which officers, other than patrol, will receive cameras. Officers working specialized units such as SWAT, school resource officers, community officers, traffic safety, and crash team officers all have unique missions within your agency. Your agency will have to decide which of these specialized units should receive BWCs. If you choose to equip your SWAT officers with BWCs, it would be more beneficial to describe their use in Standard Operational Procedures (SOP) rather than policy. The same is true for school resource officers (SRO), their use of BWCs has very specific applications and you will need to work with school administrators so that expectations are clear to both agencies. When issuing BWCs to the various officers in your agency, clear guidelines must be communicated to the officers whether it is in the form of policy or SOPs.

B. Training and Best Practices

Police agencies must have effective training guidelines for their BWC program. This training should provide instruction on how to use the equipment, how to manage video, training on your policy, and best practices from around the country. Most BWC vendors will provide train-the-trainer classes to your agency when you purchase their products. There is usually some cost associated with this training, so it is important to have your training needs written into the contract with the BWC vendor. The training you provide to your officers should give clear guidance on when to record citizen contacts, and when not to record. Best practices from around the country suggest recording anytime the police officer is taking enforcement action or dealing with a confrontational or argumentative citizen. The training should also contain situations when the officer shall not record such as in bathrooms, locker rooms, or anywhere there is an expectation of privacy. Your BWC training should encourage officers to consider citizens privacy rights and concerns. Your officers should consider the needs of public safety as well as the privacy and constitutional rights of your citizens. Officers should not record crime victims, witnesses, or citizens who want to provide crime information to officers. If you do choose to record in these instances, your training and policy should allow the citizen to refuse the recordings. An effective training program that gives your officers clear direction on the use of BWCs will ensure you have a professional BWC program.

There is controversy about whether officers should be able to review their BWC video before writing reports. We must remember the video produced from wearable cameras is just another law enforcement tool that collects physical evidence. The officer's written narrative of the incident and a thorough investigation of that incident remains the core of great police work. BWC video is not intended, nor should it be used, as a replacement for the officer's written narrative or a thorough investigation. Your agency can decide whether or not you allow the officer to view videos before writing reports. Most police agencies around the country allow officers to view their BWC video anytime they choose. The key is to train your officers in a way they understand that their written report should never be changed after viewing video. The written report must contain the officer's perceptions of the incident, even if the video contradicts the officer's account of the incident. The officer should be trained to write reports that articulate their perception of the incident and include comments about what the video appears to show when reviewing the video after-the-fact.

TASER International, Inc. (TASER) provided funding and material support for a research study conducted at their headquarters in Scottsdale, Arizona. A prospective study using a convenience sample of active law enforcement officers completed several scenarios then completed a written use-of-force report from memory. The officers were then able to review the BWC video and amend their reports as they felt necessary. The study showed that reviewing the BWC video improved the officer's accuracy in their report writing. The mistakes made by officers before viewing BWC video included miscounting or mis-sequencing events during the scenario. Some other mistakes included omitting force, warnings, or compliance during those scenarios. This study shows the importance of allowing the officer to view their BWC video before writing reports (Dawes, Heegaard, Brave, Paetow, Westin, & Ho, 2015).

The ACLU of Virginia has quite the different view about allowing officers to view BWC video before writing reports. They listed 15 recommendations to include in BWC policy for all law enforcement in Virginia in an attempt to provide consistency. Recommendation #13 reads, "BWC policies must prohibit officers from reviewing BWC video before writing their reports of incidents in which they are involved that involve use of force or alleged misconduct" (ACLU, 2015, p. 25). The ACLU says that allowing officers to review this video will undermine the fairness of review or investigatory process. They further say that allowing officers to review video before report writing could harm police-community trust. It is important for your agency to understand this controversy and create policy based on the wants and needs of your police-community relationships.

C. Managing Video Data and Storage

Managing your video data and the storage of your videos will be an ongoing, tedious, and constantly growing process. Whether you choose to use cloud storage or local storage for your video, it must still be managed by your BWC team. Ideally, your agency should have one person, for every 100 BWCs, to manage the videos produced by the BWCs. The daily activities of your video manager are to ensure the videos are properly tagged with the metadata. Metadata is simply data that can be attached to the video, by the officer. Some examples of metadata are report numbers, adding a title to the video, and adding a category code to the video. Agencies can create as many category codes as necessary and some category examples are; evidentiary, non-evidentiary, complaints, and use of force. Most video management systems will allow you to attach your retentions times to your category codes. The BWC video manager should also be responsible for ensuring your agency is adhering to your retention schedule. They must ensure that videos are not deleted prematurely, and that videos are deleted when the retention period expires. The video manager must also ensure your agency is not at risk of going over your video storage limits. There are budgetary concerns associated with the storage of video and this must be managed effectively.

The sharing of videos will become a daily event for your officers and BWC team members. Although the responsibility of sharing video should fall to the officer, the BWC manager will also be an integral part of this sharing. BWC video captured by an officer should be assigned to that officer and only the recording officer and BWC administrators should have access to that video. Officers should not be able to view other officer's video unless shared by the officer who captured the video. You should choose a video management and storage system that is user friendly and up to date with modern technology. This is primarily found in cloud storage solutions. If you are a medium to large size law enforcement agency, you should not be copying video to disk. The advanced technology available to us today allows for the sharing of video either electronically or by way of email invitations to view video from your storage system. If you are still burning video to disk in this day and age, you are behind the technological curve. The sharing of video with your prosecutor's office will be a necessary function in response to discovery motions. During your meetings with stakeholders, that should have been done before receiving your BWCs, you should have outlined a process to share video with the prosecutor's office. One vendor of cloud storage technology, TASER (thru evidence.com), provides a free prosecutor's version of evidence.com which allows the prosecutor's office their own independent system to manage BWC video. Of course, this is only if the police agency is paying for their version of evidence.com and the police must share video for the prosecutor to have access to that video.

Some agencies using BWCs have experienced a dramatic increase in FOIA requests and other agencies have experienced hardly any increase in FOIA requests after deploying BWCs. Since this is an unknown factor when deciding to deploy BWCs, your agency must consider the possibility of an increase in FOIA requests and how you plan to manage those requests. It is a good idea to meet with your county or city attorney to receive guidance on processing these requests for BWC video. Some questions that often come up are about redacting certain audio or video images before releasing the video in response to FOIA. The BWC and video storage vendor you choose should provide redaction tools in their software programs for you to easily accomplish these redaction tasks when needed. Law enforcement agencies will always have to deal with these FOIA requests, whether you are using BWCs or not, it is just one more thing to consider when using BWCs. On a positive note, your agency should be able to recover the costs associated with producing the information requested from these FOIA requests.

V. Maintaining Your Body Worn Camera Program
A. Maintaining the Equipment

Once you have all your BWCs deployed in the field, your BWC team will have to maintain the cameras, continuously audit the video to ensure the officers are properly tagging their videos, and manage the amount of storage your agency is using. You will also have BWCs that break or malfunction; if you have BWCs with mounting options, some of those will break or get lost. Your BWC team will have to manage the inventory control of these replacement items and return BWCs to the vendor for repairs when needed. The BWC program for your agency will continue to evolve over time, your team will have to continue the research and look at best practices from around the country. As the BWC and storage technology continues to improve, your agency should consider which upgrades will work best for your agency.

B. Evaluating Long Term Storage Solutions

During your agency's long term planning, you should consider a BWC replacement plan on a 5-year interval. Most vendors offer a one-year warranty and suggest that you replace your BWCs after five years. This will involve budgetary planning to ensure you have the funding for the replacement schedule your agency develops. Most BWC vendors offer extended warranties and/or replacement plans from which you can choose. Your agency should look at both of these options to see what best suits your agency. Some agencies, for fiscal planning purposes, prefer to have a defined replacement plan with fixed payments allocated over a five-year period. This makes the budgetary planning easier for the agency. Your long term planning should also include storage options for your BWC video. We have evolved from the mainframe computers, to personal computers, and now to cloud technology, this too will evolve over time. Your agency should keep up with these technology trends and continue to research and evaluate storage options for your agency.

VI. Conclusion

A. Keeping up with Technology

BWC technology has been thrust to the forefront of policing in America. This technology does have the ability to establish and maintain trust between the police and the citizens they serve, but as with any technology, there are limitations. Police agencies must be able to explain these limitations to the citizens, media, judges, attorneys, and juries. The work you put in on the leading edge of your BWC program, with regard to educating your citizens, will pay huge dividends once your BWCs are deployed in the field. The research on BWC and storage technology must be a continuous process to make your BWC program effective. Your agency should continue to keep the internal and external stakeholders informed of the status, progress, and costs associated with your BWC program. Having an effective policy in place, where the officers know what is expected of them, will ensure your program remains effective. Any time your policy is updated, you should ensure those stakeholders mentioned earlier are informed of those policy changes.

B. Ongoing Research

Your agency should thoroughly test and evaluate as many of the BWCs on the market as you can and choose the one that bust suits your agency's needs. Work closely with your purchasing department to ensure you are complying with the purchasing laws of your state. Choose a dedicated team with dynamic leadership characteristics to manage your BWC program. The correct team will make your BWC program successful. Your agency should have a BWC policy in place before placing BWCs in the field. The most innovative agencies will ensure their internal and external stakeholders are included in the policy creation. The BWC and storage technology will inevitably evolve over time, your BWC team must keep up with the research and best practices from around the country so your agency remains on the leading edge of new technology. If your agency can continue their forward thinking and evolve with the new technology, then your agency will have an effective body worn camera program. This will allow your agency to reach the ultimate goal of building and maintaining police-citizen trust within your community.

References

Alpert, G.P. (2009). Interpreting police use of force and the construction of reality. Criminology & public policy, 8(1), 111.

Alpert, G.P., Dunham, R.G. (2010). Policy and training recommendations related to police use of ceds: Overview of findings from a comprehensive national study. Police quarterly, 13(3), 235.

Alpert, G.P., & Smith, W.C. (1994). How Reasonable Is the Reasonable Man: Police and Excessive Force. Journal of Criminal Law and Criminology 85(2). 481.

American Civil Liberties Union of Virginia (2015). Model Policy.

Ariel, B., Farrar, W., Sutherland, A. (2015). The Effect of Police Body-Worn Cameras on Use of Force and Citizens' Complaints Against the Police: A Randomized Controlled Trial. *Journal of Quantitative Criminology, 31*(3), 509-535. doi: 10.1007/s10940-014-9236-3

Beckett, K., & Sasson, T. (2000). *The politics of injustice: Crime and punishment in America.* Thousand Oaks, CA: Pine Forge Press.

Belur, J. (2009). Police use of deadly force: Police perceptions of a culture of approval. Journal of contemporary criminal justice, 25(2), 237.

Bolton, B. (2015). What you need to know about body-worn cameras: A primer. *Law Enforcement Technology, 42*(1), 14-17. Retrieved from http://search.proquest.com/docview/1688607194?accountid=14665

Buhrmaster, S. (2015). 10 limitations of Body-Worn Cameras you need to know for your protection. Retrieved directly from S. Buhrmaster via email.

Business Dictionary. (2013). Retrieved from
http://www.businessdictionary.com/definition/effectiveness.html#ixzz2PGsPkXcp
http://www.businessdictionary.com/definition/efficiency.html#ixzz2gbuf79gE

Carroll, R. (2013). The Guardian. Body cameras worn by police in Rialto, California have resulted in better policing. Retrieved from http://www.theguardian.com/world/2013/nov/04/california-police-body-cameras-cuts-violence-complaints-rialto.

Dawes, D., Heegaard, W., Brave, M., Paetow, G., Westin, B., & Ho, J. (2015). Body-Worn Cameras Improve Law Enforcement Officer Report Writing Accuracy. *The Journal of Law Enforcement, 4*(6). Retrieved from http://www.jghcs.info/index.php/l/article/view/410

Dietzen, L. (2015). Body-worn cameras for law enforcement officers. *Law & Order, 63*(6), 58-60. Retrieved from http://search.proquest.com/docview/1693339081?accountid=14665

Drover, P., & Ariel, B. (2015). Leading an Experiment in Police Body-Worn Video Cameras. *International Criminal Justice Review, 25*(1), 80-97. doi: 10.1177/1057567715574374

Engel, R.S. (2008). Revisiting critical issues in police use-of-force research. Criminology & public policy, 7(4), 557.

Flanagan, T.J., & Vaughn, M.S. (1995). *Public opinion about police abuse of force.* In W.A. Geller & H. Toch (Eds.), *And justice for all: Understanding and controlling police abuse of force* (pp. 113–132). Washington, DC: Police Executive Research Forum.

Ford, T. (1997). Effects of stereotypical television portrayals of African-Americans on person perception. *Social Psychology Quarterly, 60*, 266–278.

Garrison, R. (2015). BODY-WORN CAMERAS...THE FUTURE IS NOW. *Law & Order, 63*(2), 34-36. Retrieved from http://search.proquest.com/docview/1661723851?accountid=14665

Garrison, R. (2015). The costs and benefits of body-worn cameras. *Law & Order, 63*(7), 58-59. Retrieved from http://search.proquest.com/docview/1704386574?accountid=14665

Geoghegan, S. (2015). STRATEGIES FOR BODY-WORN CAMERAS. *Law & Order, 63*(2), 30-32. Retrieved from http://search.proquest.com/docview/1661723752?accountid=14665

Graham v. Connor 490 U.S. 386 (1989).

Hoon, L. (2010). An examination of police use of force utilizing police training and neighborhood contextual factors: A multilevel analysis. Policing: An International Journal of Police Strategies & Management, 33(4), 681.

Hughes, N. (2014). Apple's iPhone led 2013 US consumer smartphone sales with 45% share – NPD. *appleinsider.* Retrieved from http://appleinsider.com/articles/14/02/20/apples-iphone-led-2013-us-consumer-smartphone-sales-with-45-share---npd

Jefferis, E., Butcher, F., & Hanley, D. (2011). Measuring perceptions of police use of force. Police practice & research, 12(1), 81-96.

Jefferis, E., & Kaminski, R. (1997). The effect of a videotaped arrest on public perceptions of police use of force. Journal of criminal justice, 25(5), 381.

Jennings, W. G., Fridell, L. A., & Lynch, M. D. (2014). Cops and cameras: Officer perceptions of the use of body-worn cameras in law enforcement. *Journal of Criminal Justice, 42*(6), 549-556. doi: 10.1016./j.jcrimjus.2014.09.008

Klinger, D. (2005). Social Theory and the Street Cop: The Case of Deadly Force. Police Foundation, Ideas in American Policing, 6(7), 1-16.

Kundra, V. (2011). White House – U.S. Chief Information Officer. Federal Cloud Computing
Strategy. Retrieved from https://www.google.com/webhp?sourceid=chrome-
instant&ion=1&espv=2&ie=UTF-
8#q=cloud+first+technology+recommended+by+the+white+house

Lee, H., Jang, H., Yun, I., Lim, H., & Tushaus, D. (2010), "An examination of police use of
force utilizing police training and neighborhood contextual factors", Policing: An
International Journal of Police Strategies & Management, Vol. 33 Iss 4 pp. 681 – 702

Lee, H., Vaughn, M.S., & Lim, H. (2014). The impact of neighborhood crime levels on police
use of force: An examination at micro and meso levels. Journal of criminal justice, 42(6),
491.

Lewinski, W.J., Hudson, W.B., & Dysterheft, J.L. (2014). Police Officer Reaction Time to Start
and Stop Shooting: The Influence of Decision-Making and Pattern Recognition. Law
enforcement executive forum, Vol. 14 Iss 2.

Maxson, C., Hennigan, K., & Sloane, D.C. (2003). Factors that influence public opinion of the
police. *National Institute of Justice: Research for practice.* Washington, DC: National
Institute of Justice.

Miller, L., Toliver, J. (2014). Police Executive Research Forum (PERF). *Implementing a Body-
Worn Camera Program: Recommendations and Lessons Learned.* Washington, DC:
Office of Community Oriented Policing Services.

Paoline III, E.A., Terrill, W. (2011). Police use of force: Varying perspectives. Journal of crime
& justice, 34(3), 159.

Phillips, S.W., Sobol, J.J. (2011). Police attitudes about the use of unnecessary force: An
ecological examination. Journal of police and criminal psychology, 26(1), 47.

Police Executive Research Forum (2011). Electronic control weapon guidelines.

Ponder, R. D. (2001). The leaders guide. Oregon: Oasis Press.

President's Task Force on 21st Century Policing. 2015. *Interim Report of the President's Task
Force on 21st Century Policing.* Washington, DC: Office of Community Oriented
Policing Services.

Roberts, J.V. (1992). Public opinion, crime and criminal justice. *Crime and Justice, 16,* 99–180.

Rojek, J., Alpert, G.P., & Smith, H.P. (2012). Examining officer and citizen accounts of police
use-of-force incidents. Crime and delinquency, 58(2), 301-327.

Rosenbaum, D., Lawrence, D., Hartnett, S., McDevitt, J., & Posick, C. (2015). Measuring
procedural justice and legitimacy at the local level: the police-community interaction
survey. *Journal of Experimential Criminology, 11*(3), 335-366. doi: 10.1007/s11292-015-
9228-9

Sigler, R.T., & Curry, B. (1992). Handling the hostage taker: Public perception of the use of force by the police. *American Journal of Police, 11*, 113–125.

Stanley, J. (2013). ACLU Senior Policy Analyst. White paper; Police Body Mounted Cameras: With Right Policies in Place, a Win For All.

Sykes, G., & Matza, D. (1957). Techniques of neutralization: A theory of delinquency. *American Sociological Review, 22*, 664-670.

Terrill, W., Reisig, M.D. (2003). Neighborhood context and police use of force. The journal of research in crime and delinquency, 40(3), 291.

Tuch, S.A., & Weitzer, R. (1997). The pools-trends: Racial differences in attitudes toward the police. *Public Opinion Quarterly, 61*, 642–663.

Vlahos, J. (2009, October 1). Surveillance society: New high-tech cameras are watching you. *Popular Mechanics*. Retrieved from http://www.popularmechanics.com/technology/military/4236865

Weitzer, R. (2002). Incidents of police misconduct and public opinion. *Journal of Criminal Justice, 30*, 397–408.

White, M.D., & Klinger, D. (2012). Contagious fire? An empirical assessment of the problem of multi-shooter, multi-shot deadly force incidents in police work. Crime and delinquency, 58(2), 196.

White, M. D. (2014). *Police Officer Body-Worn Cameras: Assessing the Evidence*. Washington, DC: Office of Community Oriented Policing Services. Retrieved from https://www.ojpdiagnosticcenter.org/sites/default/files/spotlight/download/Police%20Offi cer%20Body-Worn%20Cameras.pdf.